EDSON SEABRA JUNIOR

HOW I THINK THE UNIVERSE

A CREATIONIST MODEL FOR THE COSMO

"...and God said, create light and there was light, and separated it from darkness, creating day and night..."

GENESIS

For all those who believe that the Universe had a creator who shaped it with rigid norms for its perfect evolution and functioning.

SUMMARY

PREFACE 5

1 Universe and Space 8
2 Gravity ….………………………………………… 19
3 Common Matter and the Celestial Bodies 28
4 Solar system 36
5 Time and Light 45

PARADIGMS OF THIS MODEL 56

PREFACE

The author found in the school study of physics and astronomy a great curiosity about the origin and functioning of the universe and was dedicated to know the remote and new philosophical and scientific theories on its creation and evolution. With the advent of modern media like diagrammations and computational animations, he found his best barn of the discoveries and models that astronomers progressively formulated.

The increasing knowledge of this science thanks to the technological contribution to the art of observing and measuring the movement of celestial bodies and physical events such as gravity, black holes, dark matter, time, space and others, many gaps were left to the author on the models presented for explain the creation and functioning of the universe, since for him it did not satisfy the whole observed. The author thus began to imagine a theoretical model that explained in harmony the whole already observed and measured in astronomy and astrophysics and what is still considered a mystery.

This book *How I think the Universe* is the conclusion that came to a model that would welcome the observed and contemplate a reasonable explanation for the birth and evolution of the Cosmos, having as a fundamental in the formation of this thought a conjunction of scientific principles and creationism. The goal is to lead the reader to think of other models for the cosmos than the existing orthodox of the "Big Bang" and, perhaps, to create a new paradigm for the understanding of this great mystery and spectacular existence that we call the Universe.

We must take into account during the reading of this book, the creationist and scientific philosophical content of this model, without which the meaning of the whole is impaired.

Universe and Space

We learn on school benches that space is a vacuum, a void where the stars and the firmament are contained and everything else observable in the sky. However, tracing back to the past, we observed that in ancient Greece there existed the concept of space being filled by a discrete fluid which they called Ether where all the celestial bodies were immersed. This concept appeared in the works of Pythagoras and accepted by the Greek philosophers; Socrates, Aristotle and Plato. It had a long life, having explained the movement of the celestial bodies around others as a result of the whirlwind that they caused in this Ether as they moved.

Questions about the existence of Ether only came to light with Newton's gravitational theory where the relations between the celestial bodies were due to a mysterious force of attraction at a distance between the massive bodies regardless of a physical medium to connect them. Later on with Einstein's theory of relativity, the old and respected Ether is now seen as the space-time fabric itself with new properties. although all these models competing with the Ether primate, this still finds survival no longer as a timely physical means of filling the void between the stars, but as Reich (40 and 50 of the twentieth century) defined an ocean of cosmic energy whose properties is that they determine actions at a distance such as gravity.

More recently in the second half of the twentieth century, NASA's project aimed to detect whether Earth's spinning motion twisted the space / time fabric, was placed stationary in orbit in space four gyroscopes pointing to a distant star and observed a small but measurable misalignment of this aim in the observed time proving that the fabric space / time is twisted as a consequence of the rotation movement of the

Earth and therefore it is constituted of some massive substance.

For all that has historically been spoken of what space is and added the modern discoveries of some of its physical properties as; torsion, expansion and compression already observed by the modern science of astronomy, suggests a model of what will be *the space* and thus we construct it as; *a highly massive substance with high fluidity, compressibility, expandability and mobility, formed by an enormous amount of very small particles that we call "primordial", codified and indivisible, part carrying greater energy and mass and others of lower energy and mass with the property of organization in various forms and vibrating at the speed of light.*

It is from the space that originates and returns everything that we observe in the universe what for so much we define it: *the Universe in our creationist model is a closed system, consisting of a finite continent with constant and three-*

dimensional volume that holds as the only content the substance space as previously defined.

We have the Universe as a creation of a superior intelligence that conceived and programmed it within a constructive and autonomous logic with primordial particles encoded, provided with volume, energy and mass necessary for the attainment of the architect. These particles can be understood as the genome of the Cosmos that establishes all its form of behavior, functioning and manifestation. We can figuratively see these primordial particles coded as the letters of our alphabet where, by combining, we can name objects, numbers, actions, ideas, and all human knowledge, just like the computer combining sets of bits that assume only two conditions; switched on and off may represent electronically any such knowledge.

This is the starting point of this creationist vision of the Universe as we know it, observe it, measure it, feel it, and whatever else we discover about it. Thus space is an intelligent entity made up of pre-coded particles with mass and energy to assume a multitude of combinations and specific objectives, but conditioned to a pre-established logic that guides the whole organization of that system. These particles can thus be organized in different forms: *the pattern* (original form, total organization), constructed of the primordial particles distributed homogeneously and that constituted the initial state of the universe and still its greater volume; as *dark matter* that are places of space where a large volume of the primordial particles of low energy and mass are continually compressed by the others and become denser, beginning the construction of a reorganized subsystem of the space "dark matter", occupying the second largest volume of the manifestation of space in the universe; *black holes*, third largest volume of space manifestation, are places where dark matter undergoes a high degree of compression and density causing in the primordial particles at its center an enormous reduction of its degree of freedom of vibration, transforming almost all its kinetic energy into potential, initiating a process of connection between them that selectively and obediently to their primordial codes, create the fundamental composite particles of common matter as we know it; protons, electrons and neutrons the basic components of the atoms that will by fusion process form the natural chemical elements of the Periodic Table of the Chemical Elements and constituting in the smaller volume of the forms of manifestation of the space.

These newly created composite particles bring with them new and complex codifications resulting from the combinations of primordial particle codes that integrate them. They in turn have greater density than the black hole that created them, because it is the product of the limit of the densification of the primordial particles that created them. The same is not true for the atoms and chemical elements and the substances resulting from their combinations, since they keep in the formation of their structures of existence lower density, by the greater distance between the fundamental composite particles that form them.

The primordial particles of the rest of the space permeate the atoms filling the places not occupied by their nuclei and electrons, since already formed by them, do not overlap each other.

THE UNIVERSE AND THE MAIN FORMS IN WHICH THE
SPACE SUBSTANCE PRESENTS

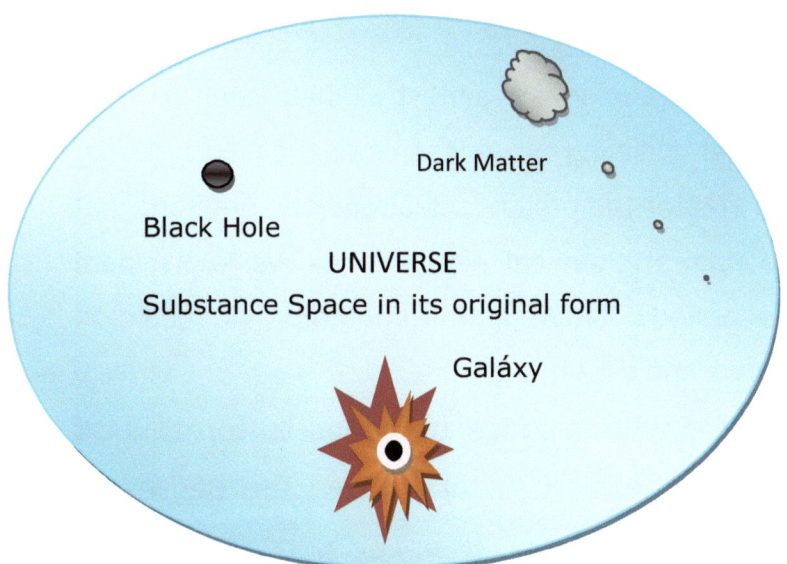

Substance Space in its original form, consisting of primordial particles encoded of high and low energy distributed harmoniously and that constitutes still the greater volume of the space.

Dark Matter, are places in space where the lower energy particles are compressed by those of higher energy, increasing the density of space in these places. A localized process of pressure (gravity) is formed by the rest of the space on part of itself, creating the second largest form of space organization.

Black hole, where the highly dense and massive dark matter suffers at its center a critical compression that the degree of freedom of primordial particles of lower energy tends to zero, transforming almost all their kinetic energy into potential, initiating a process of connection between them that creates the protons, electrons and neutrons that form the natural chemical elements as we know them from the periodic table, and from them, the common matter as we know it.

In places where space proceeds to form "dark matter" and these in "black holes," the degree of freedom of mobility of their primordial particles diminishes and proportionately transforms most of their kinetic energy into potential energy. At the limit of this concentration in the black holes, when the composite particles "Protons, Electrons and Neutrons" are created, they are then ejected axially through their axis of rotation and orbiting it. Neutrons, a nuclear fusion of a proton and an electron, are highly conductive particles, devoid of electrical charge, essential to stabilize the nuclei of the atoms, functioning as an electrical connection between the protons in their nucleus, making them function as a single electrical charge, avoiding the repulsion between them.

It is thus created under mastery and architecture, the basal structures and bricks for the construction of matter as we know and admire it, obedient to the rigid norms of creation contained in the primordial particles that creates everything, pervades everything and orchestrates the cosmic events of which we are a product and users. We know very little about this, but we have been given the ability to think, feel, observe, measure and reason what leads us to be irreverent and arrogant to philosophize and model the creation and evolution of the Universe, in the eternal quest to understand the origin of our existence. As the primordial particles in the form of dark matter, black holes and fundamental compound particles occupy less volume of the universe than it occupied in its original state, the rest of the space still in its standard form expands, since it has more and more volume of the Universe , and pushes with this expansion, these conglomerates of subsystems. This expansion will stop one day and the Cosmos will stabilize, for as these conglomerates are formed and the rest of the space expands, its original density decreases, and with it its capacity to generate new conglomerates. With this, there will come a time when the forces of these forms of space organization will be balanced, keeping only a few exchanges between them, but without forming new conglomerates. In this process, the forces of gravity on each of these conglomerates also undergoes a reduction, and in some cases of dark matter still incipient, decompress some of its primordial surface particles to its original conditions. We perceive in this plot that Universe and Space are inseparable and are confused in our perception, because they keep a relation of sense between itself, is the continent by the content and sees versa, a metonymy. Formed by primordial particles, Space has a discrete nature even though it seems continuous to us by the tiny size of the particles that form it, although its discrete nature functions as a single entity with an intelligence, regardless of its size and presentation. We say here that what occurs in a seemingly isolated form is part interrelated to the whole for its continuous reorganization into an immediate cause and effect principle that guarantees its existential and functional stability.

Gravity

Gravity is the continuous pressure resulting from the compression of the rest of the primordial particles of space over those places where they have lost their homogeneous distribution; "Dark matter, black holes, protons, neutrons, electrons, atoms, and ordinary matter as we know it." It arises on the created subsystems of the harmonic organization of primordial particles. This compression that we call gravitational force is perennial and without loss of energy, because it results from the existence of space, but with the property of diminishing its force with the distance of the layers of compression of the rest of the space on these subsystems like what happens in the effect crowd.

It is a relentless force, localized and arising from the effort of the rest of space to occupy the place where these subsystems arise in the

Cosmos, that is, it is directly proportional to the quantity of primordial particles existing in each type of subsystems and reduces rapidly with the distance of them, in the velocity proportional to the inverse of the square of these distances as already calculated by Isaac Newton.

Like the volume occupied by the primordial particles densified in the reorganized subsystems of space; dark matter, black holes, composite fundamental particles, and ordinary matter are smaller than the volume they formerly occupied in their standard state, the rest of the primordial particle space expands, since it has more volume in the universe to occupy and causes an inevitable decrease both gravity positive and negative. This continuous expansion generates a compression of the rest of space on these subsystems that collectively affect them differently when near and far from each other, so that; when a set of these places is sufficiently close to each other, that is, there is intercession of their gravitational fields, there also acts on this set a compressive force that surrounds them trying to unify them, and when they are far enough, that is, there is no intercession of their gravitational fields now arises a distension of the space between them that tends to move them away because the rest of space now expands between them.

The gravity that acts on adjacent subsystems does for joining these places, but when they are sufficiently distant it does to push them away. A negative gravity emerges after a point of inflection *S* between them that depends on the masses involved between a subsystem/set of them of another/others of the space, to be seen as near/distant. Distances greater than this point of inflection between places of space causes them to suffer only the local gravitational pressures, and now between them, an inverse pressure arises, which will separate one from the other.

GRAPH OF THE PERFORMANCE OF THE REST OF SPACE ON SUBSYSTEMS ARISING FROM IT

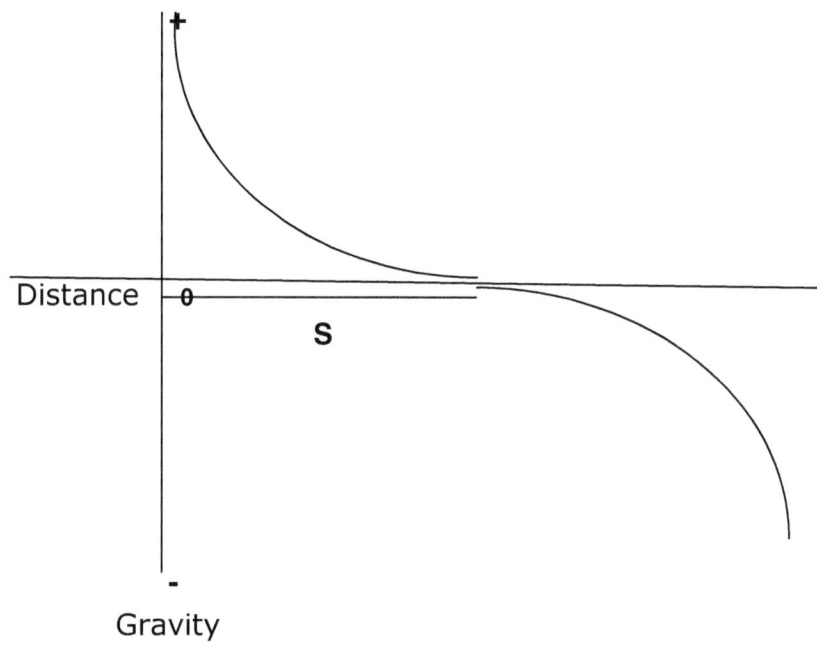

Thus we can say figuratively that we have attractive gravity when the compression of the rest of the space tends to unite subsystems of the space next to each other and repulsive gravity when it tends to move away subsystems of the space distant from each other. The space has several properties, where the non-distributivity of the compressibility is the main responsible for the decrease of the force of gravity on the subsystems when it moves away.

Thus the rest of the space is compressed more on the surface of the subsystem and successively decompresses as it moves away from it, ie it is as if compressing a spring, the pressure does not distribute equally in all of it as happens in a helical spring of compression but rather that it would be larger in the opposite spiral where the force is being printed and decreasing in successive spirals as it approaches the end where the force is being applied. It would be like the crowd effect, the person who is ahead suffers the push of all that are behind him and so successively diminishing each person after another.

Thus when space forms a derivative subsystem, it undergoes continuous pressure from the rest of the space proportional to the original quantity of matter of space contained in it and decreasing with the square of the distance of its surface.

Let's look at an example of this space behavior; suppose a cube of volume v hermetically closed of 50 cm of edge with air at a pressure p and now a ball of bowling appears in its center, reducing to v ', the volume now available for the air. When the air pressure is measuring again, we will find a pressure $p' > p$, because the volume now available for air v', is less than v, and air will be again homogeneously redistributed in this new volume v' and will reach a pressure p' but equal in any place measured within this new volume v' as a function of the distributive property of the gas pressure. Now imagine that we have a device that can measure the density of space within that cube of volume v and find a density of space d at any point in it before the existence of the bowling ball. Now with the bowling ball in the center of the cube and a remaining volume $v' < v$, we make several measurements of the space density in the cube from the surface of the ball and successively from it moving away, we will find on the surface of the ball a density **d'** and in the following measurements, densities progressively smaller in such a way that **d' > d'' > d '''> ...> d**, that is, space is not homogeneously redistributed in this new volume **v'**, but rather presents a higher density and pressure on the surface of the ball and is continually reduced in the subsequent layers, this is due to the non-distributive property of the compression of space, without which there would be no gravity nor matter as we know it.

THE CUBE WITH THE BOLICHE BALL IN ITS INTERIOR

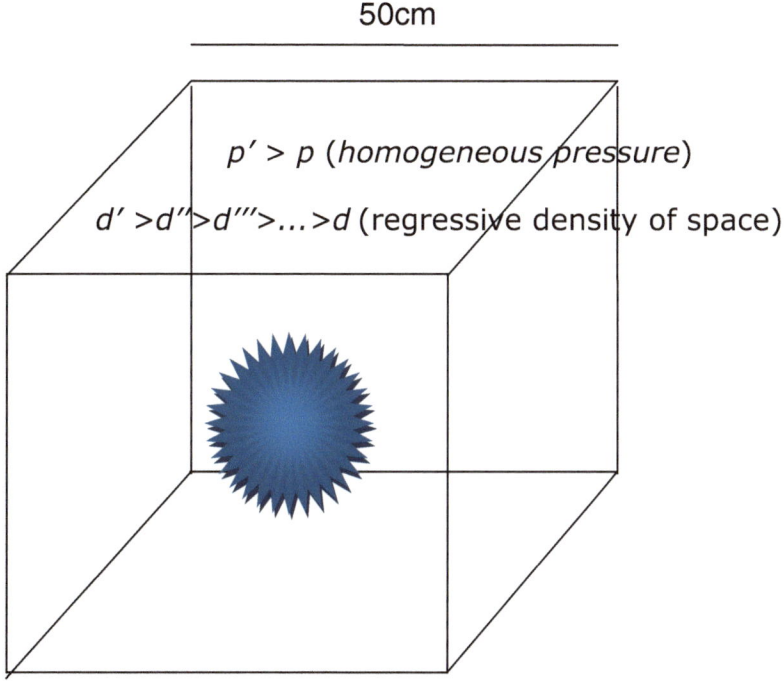

This property of the space that creates gravity is also responsible for the generic spherical shape of the celestial bodies, because this is the geometric solid that presents the largest volume per surface, thus, the space concentrates its compression on the smallest surface to a given volume of matter and reaches the highest possible pressure for the same effort and consequently the greatest gravitational force, since it acts on a smaller area. Gravity is thus a consequence of the existence of space, its property of contracting, expanding and modifying its configuration. It is the greatest force and primate in the cosmos; indestructible, continuous, perennial and universal.

The existence of a gravitational force in a place of the space indicates that a deformation occurred there by densification, and the larger the force, the greater its density and vice versa. The greatest density and therefore the greatest gravity per unit mass is in the fundamental particles "Protons, Electrons and Neutrons" followed by black holes, dark matter and to a lesser extent in common matter due to the degradation of their concentrations of primordial particles.

Common matter and celestial bodies

Common matter is formed by a large scale of atoms formed by the combination of the fundamental particles; protons, neutrons and electrons which are created by the fusion of primordial particles of low energy and mass compressed inside a black hole and ejected axially to its axis of rotation and orbit it by force of the vortex of the space in its environment. These fundamental particles of the atom now subjected to new forces external to the black hole that created them combine to form in a fusion process the atoms as we know them, first the hydrogen and then the others in a continuous process of volume growth and densification.

As the lighter chemical elements, notably Hydrogen and Helium, are formed around the black hole, the formation of clouds begins, which, by the action of the pressure and rotation of the space around it, form heavier elements that are agglomerating and in combination they create diverse substances on a large scale, and in sufficient quantity and time they will form nebulae, stars, planets, moons, cosmic dust, comets, asteroids, and all visible matter in a galaxy. Thus, a galaxy is formed from a black hole that will constitute its future center. The formation of all visible matter in a galaxy like the "Milky Way" requires an enormous amount of primordial particles concentrated in a great black hole that gave rise to it. Just as not all dark matter will have enough mass to turn into a black hole, not all will have enough primordial particles to be source of a Milky Way galaxy. Although smaller black holes can create the fundamental particles of the atoms and some lighter chemical elements, only those of great volume that reach a mass, density and high gravitational field, are able to generate large conglomerates of common matter and form a galaxy. The movement of rotation of the galaxies and of rotation and translation of all the matter contained therein are indispensable for their grouping and formation of their celestial bodies. Mobility is a natural state of space, everything that exists in it is in constant movement, it is a property of its existence.

The celestial bodies in a galaxy are formed in a process of accumulation of chemical elements that as they collide and combine, form more complex and dense substances, which under high pressure and temperature are aggregating and forming the common matter as we know . The centrifugal force generated by the velocity of the vortex caused in space by the rotating black hole in the center of the galaxy is in charge of organizing this material that progressively growing in mass and moving away from the center of the galaxy are consolidating in several types of celestial bodies depending on the mass and what kind of substance they have. An important property of space for this to happen is the torsion resulting from the trailing of its particles by a massive body in rotation due to the frictional forces between them and between them and the body that rotates, forming a three-dimensional vortex of enormous amplitude that places in movement of translation and rotation all matter immersed in it. Since this friction is not perfect, and the density of space that assists that force and causes gravity, decreases as it moves away from the black hole in the center of the galaxy, this velocity of the vortex of space decreases toward its periphery until it ceases.

In this way the translation movement of the celestial bodies results from the thrust upon them of the space in rotation around the center of the galaxy and its rotation movement of the difference between the greater thrust of the space on the tangent of the side of the star closest to the center of the galaxy. galaxy and smaller on the tangent from its far side. Thus the closer the star is to the center of the galaxy and the greater its diameter, the greater will be the thrust of translation and rotation. The direction of rotation of a star will always be contrary to the rotation of space that causes these movements in it and so we can say that the great stars do not move through the space in which they are immersed, they are stationary in them and who moves is the space itself which contains them and push them together and give it the rotational state.

In the following figure we observe this behavior where the vortices **V1 > V2 .... Vn > V(n + 1)** created in space by the black hole of the Milky Way when rotating in a clockwise direction, pushes in the same direction the Sun in its translation movement and also causes by the degradation of the velocity of the successive layers of that vortex the rotation of the Sun counterclockwise because the velocity of the vortex in the circle of the Sun **Vn** closest to the center of the galaxy is greater than in the circle **V(n + 1)** that is further away. These vortices in galaxies, stars, planets, and other large rotating stars, require time to reach their maximum uniform circular motion, as well as the movements of translation and rotation of those who suffer their effects. In fact, who determines the movements of translation and rotation of the stars is the movement of the space that contains them and how massive and bulky they are.

Milky Way, the SUN and its rotations

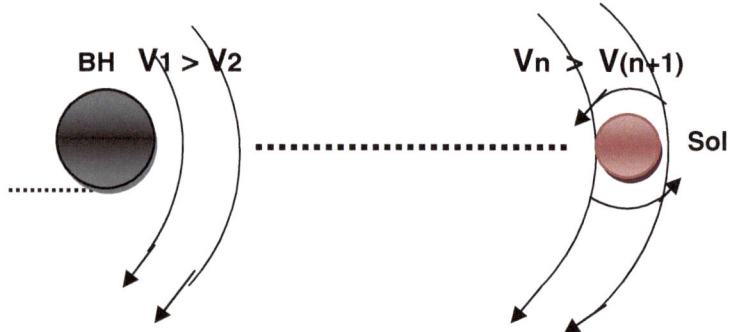

Lines of the vortex in the space of the Milky Way
BH(black hole of the Milky Way)

We see in this model that what causes the movement of translation of a celestial body around a more massive one in rotation, is the thrust of the substance space in rotation around the most massive on the less massive, the smaller the distance from that to the orbit, the higher the thrust velocity to which it will be subjected and, therefore, the greater the orbital velocity. In a similar way, the rotation movement of a less massive body in translation of a more massive one, derives from the differential of the vortex thrust of the more massive star on the less massive one as a consequence of its diameter, the larger the diameter of the orbits the larger the differential and the greater the rotational force.

These vortices, as in any other massive body in the universe, have the weakening of their gravitational field in relation to a smaller radius vortex, with the inverse of the square of the distance gradient separating them, for the following reason; the surface of a sphere of radius **r** is $S = 4\pi r^2$ and a surface of greater radius distance **dr** of the surface of radius **r** will have a surface $S' = 4\pi(r+dr)^2$ and the relation between they, $s' = s(1+dr/r)^2$. The deformation caused in space by a massive body with radius **r** gives rise to a counteraction force **f** of the rest of the space on it which depends only on the quantity of its mass that once constant, **f** will also be constant, and its pressure which causes the gravity will be $p = f / 4\pi r^2$ and for a surface of radius $r' = r + dr$ will be $p' = f / 4\pi(r+dr)^2$. Let **p'** as a function of **p** and simplifying, $p' = p / (1 + dr/r)^2$, that is, the decrease of gravity over a massive body in space, occurs with the inverse of the square of the distance gradient for its surface, which could not be different, since the spherical surfaces grow directly proportional to the square of its radius and the pressure of gravity is the result of a force divided by the surface in which it acts, then its decrease will occur inversely proportional to the growth of these successive ones areas of concentric spherical surfaces.

Solar system

Formed by the Sun a rotating star and circulating around it nine planets, moons and other smaller celestial bodies, is a stable system, where the movements of translation of the planets that compose it happen of the rotation of the space around the Sun. For the stability observed from this system we can intuit that these celestial bodies have been in interaction for a long time and that the vortex of space created by the rotation of the Sun has already stabilized.

The rotation of the sun occurs in the counterclockwise direction, which is in conformity with the hourly rotation of the galaxy, therefore in the opposite direction of rotation of the galaxy. The same is not true of the planets in relation to the sun whose rotations occur in the same counter-clockwise direction as the Sun, not keeping conformity with the rotation of the Sun but with that of the Milky Way, except for the planet Venus. We can say from this fact that the Sun and the planets were formed in different places of the galaxy, but along the same belt of the vortex in the space of the Milky Way, and later the vortex of the space created by the rotation of the Sun captured them in several moments of its already accomplished 22 laps around the center of the galaxy.

The fact that the planet Venus is the only one to have clockwise rotation contrary to the Sun, means that of the planets of the solar system it was the first to be captured by the Sun, since the anti-clockwise vortex of the Sun's space had enough time to reverse the initial counter-clockwise rotation of Venus produced by the galactic hourly vortex which consequently caused a zero rotation of Venus at some moment in its existence around the Sun. In this way a comparative study of the movements of translation and rotation of the planets of the Solar

System in relation to their time of rotation, diameters and distance to the Sun, it will be possible to establish a chronological order of their capture by the Sun during their already twenty-two trips around the center of the galaxy.

This process of capture also occurs between the planets and their natural satellites, in the case of planet Earth the rotation of the Moon is still in the same direction counterclockwise of the Earth, but at a very slow rotation speed, bringing information that a long time that it was captured by the Earth vortex and certainly at a time prior to the capture of the set Earth / Moon by the Sun.

The following figure shows the "V" thrust forces created by the anti-clockwise rotation of the space around the Sun to which the Venus and Earth planets are subjected, and the current directions of rotation of these planets; Venus (hourly) and Earth (counterclockwise).

SET SUN, VENUS AND EARTH WITH THEIR ROTATIONS

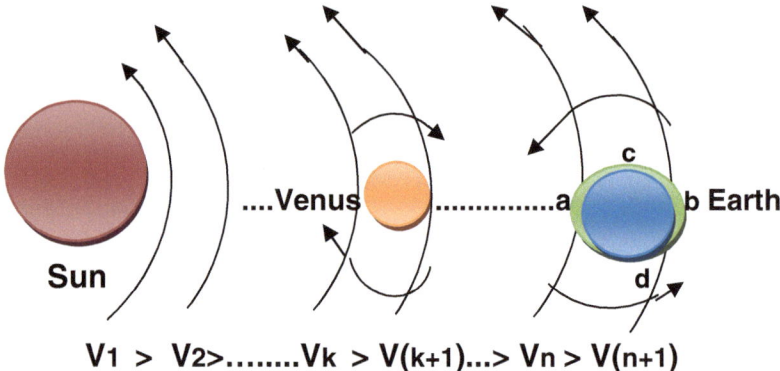

$V_1 > V_2 > .... V_k > V_{(k+1)} ... > V_n > V_{(n+1)}$

In this model the vortex of the Sun tends to rotate the Earth and the other planets clockwise opposing their current counterclockwise rotation. This thrust contrary to the current rotation of the Earth causes it to continuously decrease its speed of rotation, increasing the daytime and nighttime period to a point where it will stop and slowly begin its new rotation clockwise. This thrust of the Sun's vortex over the Earth's surface causes at points "**a**" and "**b**" in the figure, elevations on the liquid surfaces of the Earth, where at "**a**" will always be greater than at "**b**" because the thrust in the shell (**d-a-c**) is greater than the thrust on the dome (**d-b-c**) and at points "**c**" and "**d**" a depression appears on its liquid surfaces being greater in "**d**" that receives the translation thrust of the space caused by the rotation of the Sun and smaller on the "**c**" opposite side, which pushes the Earth away from the space in front of it. With the rotation and translation of the Earth, these points of the surface are changing and creating successive tidal effects.

The planets orbiting the Sun move away from and approach it respectively as they orbit the Sun in relation to the galaxy vortex thrust. As the diameter of the Sun is much larger than the diameter of the planets, the thrust it receives from the thrust of space is much larger, so when the planet in its translational motion is ahead of the Sun, that is, it receives the vortex of the galaxy before the sun, it tends to move away from the planet because the flow of space tends to flow more easily around the planets than around the sun. This distances the sun from the planet and we have aphelion there, and when planet in its movement of translation is after the Sun, that is, it receives the thrust of the galaxy vortex after the Sun, this tends to approach the planet for the same reasons that did it to distance itself in the previous situation and here we have the perihelion.

The eccentricity of the elliptical orbits of the planets varies according to their different diameters, masses, distances to the Sun and time in which they are part of the Solar System. In this model we can understand the precession of the perihelion of Mercury as the fact that its great proximity to the Sun also suffer its orbits disturbance of the swirl caused by the flow of space in regions very close to the Sun.

ELLIPTICAL MOVEMENT OF THE PLANETS

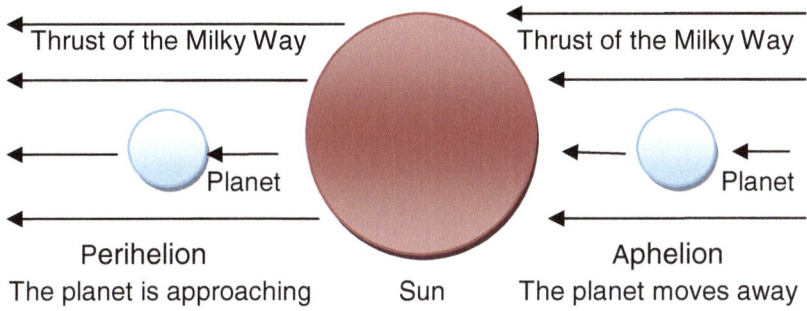

As the set Moon Earth formed at a time before they were captured by the Sun's vortex and become part of the solar system, the anti-clockwise vortex of Earth's surrounding space has had enough time to significantly decrease counterclockwise rotation of the Moon, which with the necessary time, will revert this sense of rotation to schedule keeping there the Moon in conformity with the rotation of the vortex of the Earth to which it orbits. This inversion of the Moon's rotation will occur before the rotation of the Earth turns to be hourly to conform to the anti-clockwise movement of the Vortex of the surrounding space of the Sun following the example of what happened to Venus.

SET EARTH MOON WITH ITS ROTATIONS

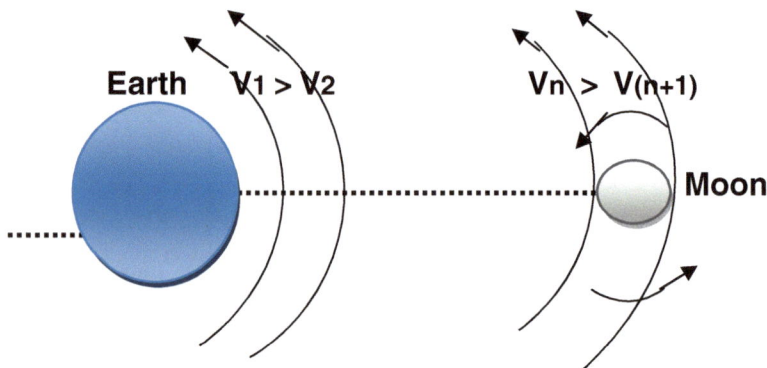

Lines of the vortex in the space of the Earth

We see in this model that the vortex in space caused by the rotation of the Earth tends to rotate the Moon clockwise opposing its current counterclockwise rotation. This contrary thrust on the Moon causes it to continuously decrease its speed of rotation to a point where it will stop and slowly begin its new rotation clockwise. Similarly, the Earth will revert its sense of rotation to clockwise by force of the anti-clockwise vortex formed in the space around the Sun and when this happens the Earth will reverse the rotation of the Moon to counterclockwise again.

Time and Light

As we said initially in this philosophical model, the Universe is formed by two unique real and concrete physical primary entities; 1 - "the continent" which is the fixed and finite three-dimensional volume that shelters the substance space as its unique content, in its various forms of presentation and 2 - "substance space" which is highly massive content with high fluidity, compressibility, expandability and mobility consisting of encoded indivisible primordial particles; part of them carrying high energy and mass, and others of lower energy and mass, vibrating the speed of light where the speed of light is the compass of the immediate change of place in space that every primordial particle continually makes in its eternal and constant process of vibration.

The time in this our concept of the universe, is the measure of the interval between one position and the next immediately next to each of the primordial particles of space in its process of vibration, time is then a property of the particle which indicates the compass in which its motion vibration is happening. It is not a fourth independent physical dimension in the universe as it is currently theorized, but a variable that changes with how much the particle is moving in a cycle, which changes when its degree of freedom of movement increases or decreases caused by the smaller or larger number of them at a point in space. Time thus reflects how much a place in space is more or is less dense of primordial particles, is the result of the relation between two properties of primordial particles; displacement and speed.

Thus, if we compare what happens with time between a place occupied by the substance space with a density "d0" which has a degree of freedom of displacement of its particles "s0" and another place occupied by the substance space compressed to a density "d1" two times greater than "d0" and if we consider for didactic effect that this relation density and degree of freedom is inversely linear, the degree of freedom of displacement in "d1" will be "s1" twice lower in "s0" and by applying in both the formula of Newton's velocity, the time "t0" and "t1" of displacement of the particles at the speed of light in "d0" and "d1" will be:

$$t0 = \frac{s0}{v0} \quad e \quad t1 = \frac{s1}{v1}$$

As $v0 = v1$ = speed of light and $s0 = 2s1$ we have:

$$\frac{2s1}{t0} = \frac{s1}{t1} \quad e \quad t1 = \frac{t0}{2}$$

We then see that the measure of time "**t1**", that is, the displacement interval of the particles in "**d1**" is 50% smaller than the displacement interval "**t0**" in "**d0**" because "**d0**" is 50% less dense than "**d1**" and its primordial particles move double "**2t1**" in each movement. Thus the measurement of the time or the displacement interval of the particle varies according to the density of the space in which it is being measured, which could not be different, because of the same velocity of change of place of the particles that constitute the substance space independently of different forms of concentration. This is because as the space is denser, less is the distance between the primordial particles and the smaller its displacement, which always occurs the speed of light, increasing its frequency " " of oscillation and decreases its period "T". As in wave physics, T = 1 / f, neither the displacement nor the frequency cannot be zero, as indeed they are not, no matter if a place in space is dense or expanded, its particles will still always be in vibrating motion. The time will then have the necessary duration to accommodate the movement of the particle in the place of the space in which it is found. This is the cause of the dilation and contraction of time which occurs simultaneously when the expansion and contraction of space.

The intensity of gravity is the result of the density of the space; places with larger gravitational field will have the measurement of time less than in places of low gravity. A moving body does not directly change the measurement of time, what happens is that in front of it a compression wave forms in the space that increases its density and decreases the degree of freedom of its particles and consequently the measurement of time as we saw above and not the speed of the body itself.

What determines the measurement of time anywhere in the universe is the state of compression or decompression of the primordial particles of space at that point.

Light arises when an energy source disturbs the space by driving a set of primordial particles surrounding that source which receives those pulses and propagates it radially to the immediately adjacent particles and so on. The energy that caused this stimulus is then transported in a wave effect until it encounters an obstacle and releases it through the movement of the primordial particles immediately adjacent to the screen. Throughout this transport there is no loss of energy because the particles involved in this transport process are already saturated with their own native energy which makes them vibrate at the speed of light and continuously keeps the space in motion generating permanent spatial waves.

The fact that light has constant velocity and is the speed limit in the Universe and its existence is already born at this speed is due to the fact that it uses the primordial particles for its propagation and manifestation in an obstacle, so its velocity does not is greater nor less than the speed of vibration of these particles themselves. When a quantum of specific energy selectively disturbs space, it propagates in waves at the speed of light that will manifest itself as luminescence in an obstacle. Light is thus the sum of a process of energy propagation that travels in waves through space and particle when it reaches an obstacle.

We can exemplify this behavior with Newton's pendulum "balancing-balls" where a kinetic energy transport from one end to another of its spheres occurs as shown in the following figure, resembling the luminous energy that uses the particles of the space for its transport and unloads it at the end of an obstacle.

BALANCING-BALLS

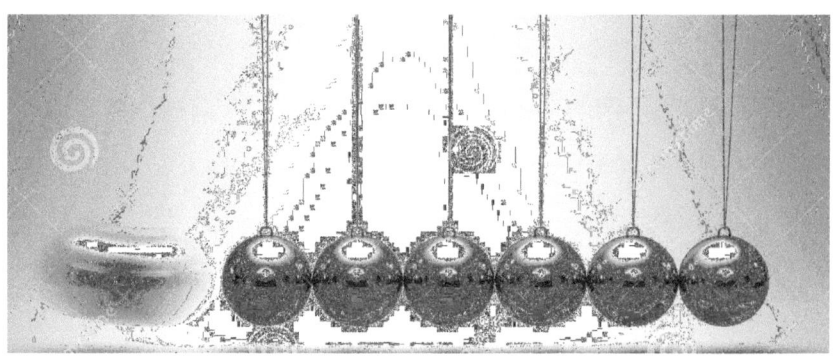

The kinetic energy of the blurred ball to the left of the photo reaches the first of the six stopped balls and propagates this energy through the others without moving them to the sixth ball to the right that does not have to whom more to transfer this energy acquires this synergy and moves . Likewise, the disturbed space carries this energy through the primordial particles until it reaches an obstacle and discharges the energy through the primordial particles immediately to the bulkhead. In this context we can intuit the duet wave + particle of light where; its rippling existence occurs only during the transport of energy from the first to the sixth ball without transport of matter, and particle when the sixth ball travels and reaches a target discharging that energy. As the transport of energy in space takes place in a discreet way, because it occurs from layers in layers of particles, it propagates by pulses of energy the speed of light that gives it a perception of being continuous.

We can observe this same behavior in the double slit experiment in the figure below; the sunlight passing through the slit **S0** of plane **A** causes a single disturbance and transport of energy in the particles of space between planes **A** and **B** and propagates harmonically. When this perturbation **S0** passes through the slits **S1** and **S2** in the plane B, they cause two perturbations to transport energy in the particles of space between planes **B** and **C** of the same intensity, independent and identical in form, but concurrent with each other, since they affect the even space between planes **B** and **C**. The interaction between these two energies generates a resultant perturbation in the space between planes **B and C** different from the originals and consequently a different resultant particle discharge in the plane **C**.

THE DOUBLE SLIT EXPERIMENT

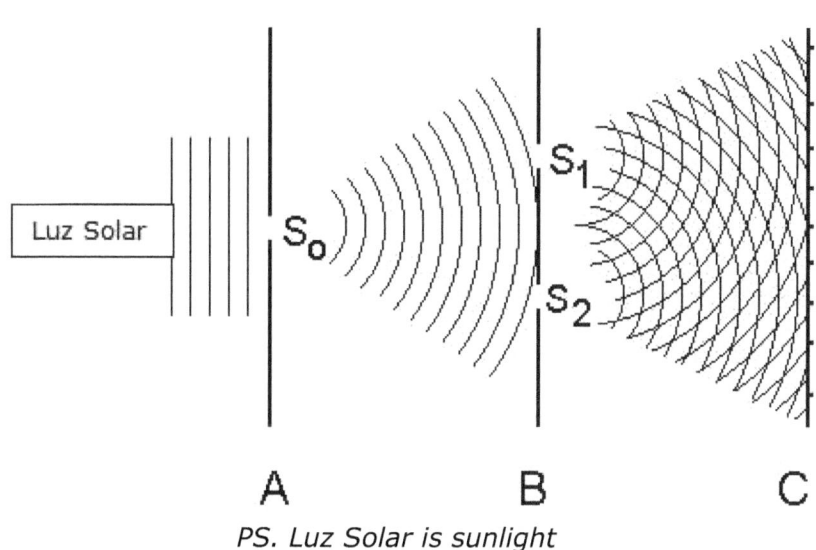

PS. Luz Solar is sunlight

When these two stimuli are in the space **BC** those that have corresponding opposing direction vectors cancel out and those that have 100% equal vectors are added together and between these extremes there will be a gradation of nullification and sum of the corresponding vectors inclined to each other. Consequently the final energy arriving at plane **C** will always be less than the sum of the stimulus energies in slits **S1** and **S2**, as there are losses by antagonistic stimuli in the path between **B** and **C**.

Paradigms of this model

Any thought, especially when it involves abstract philosophical beliefs and reasoning that although it has foundations in the culture and reality of the facts that surrounds us, gives rise to divergence with the current status quo of this theme, thus enumerating the basal paradigms that this philosophical model brings in the formation and evolution of the universe for the reader's reflection in his analysis of this reading.

1 – Space becomes a substance, pre-programmed by a superior intelligence and provided with the means necessary for the execution of the architect and ceases to be only a vacuum between the celestial bodies and without interaction with them. It is from where everything we know originates and returns, and that we discover in the future. It is a closed system, consisting of a finite, three-dimensional and fixed volume filled with particles carrying mass and energy necessary for its evolution and functioning that gives life to the Universe. The energy of each particle proves the resource for its movement and does not change with its mass, so in this model we do not have the concept that mass is transformed into energy or the inverse, what we have are changes in the energy state of particles and their concentrations. In this concept Einstein's expression $E = mc^2$ is not an equality but an equivalence, that is, the energy contained in a mass "m" corresponds to that quantity multiplied by the square of the speed of light. It does not accept the existence of the Big Bang and the concept of cosmic inflation associated with it. The cosmic microwave background radiation attributed to the "Big Bang" is in this model the vibration of the primordial particles that form the space.

 2 – Gravity is the greatest force in the universe and is no longer exclusively of union between massive objects, but also of repulsion between them, depending on their masses and the distance that separates them. The galaxies are not moving away from each other because of the expansion of the universe, but a repulsive pressure coming from the expanding space between them. The concept of gravitational waves changes into space waves once resulting from perturbations in the primordial particles that form the space. The light itself, electromagnetic waves of radio, television, cell phones, etc., are products of the perturbations by stimulus of the primordial particles of the space that is the means that provides their existences and defines their properties among which its speed that can not be superior to the velocity of vibration of the primordial particles themselves which is with the speed of light.

3 – The protons, neutrons and electrons are formed in black holes by the fusion of primordial particles encoded from space and carry with them codes of behavior for the formation of atoms, unitary elements for the formation of matter as we know it. Thus matter as we know it and everything else it provides, including the existence of plant and animal life, is not the work of chance, but a controlled process of creation, evolution, and functioning. The celestial bodies of a galaxy are the result of the agglomeration of the material formed around the black hole that originated and as they increase in volume and mass, give rise to these bodies that in great quantity will form a galaxy like the Milky Way with a black hole in its center.

4 – The solar system becomes a tiny set of celestial bodies pre-existing in a belt of the Milky Way, formed by a star "Sun" that constitutes its center and orbited by planets, moons and other smaller bodies captured by the vortex of the Sun and they pass to depend on it; their orbits, rotations and environment. The tides on Earth are the result of the pressure on the liquid surfaces of the Earth of the rotational and impulsive force of the Sun vortex that gives the Earth its movement of translation and rotation and not gravitational product of the Moon on the earth that in fact are of repulsion and not of attraction.

5 – The elliptical orbit of the planets occurs because of the greater force of the galaxy's vortex over the Sun than over the Planets, due to the greater flow of the galaxy's vortex over smaller celestial bodies than over larger ones. So when the thrust strikes the planet before it reaches the Sun, the Sun moves away from the planet and we have the aphelion there, and when we have the thrust striking the planet after reaching the Sun the Sun approaches the planet and there we have the perihelion . In the aphelion the translational thrust on the planet around the Sun is smaller and its orbital velocity decreases and in the perihelion the translation thrust on the planet around the Sun is greater and its orbital speed increases, as foreseen in the first Law of Kepler. If at the point of aphelion a planet stopped orbiting the Sun, it would gradually move away from the planet and in turn that would end up leaving the solar system, and if the planet stopped orbiting the Sun in the perihelion it would gradually approach the planet and absorb it . The movements of translation and rotation of the celestial bodies around one another are due to the vortex in the space created by the rotation of the orbiting star that pushes them and gives them these movements and not due to the force of gravity between them that is purely axial, nor due to the deformation of space / time of Einstein that although it exists in the surroundings of all matter is not its placeholder, because its orbits are beyond this deformation space / time.

6 – The time in this our concept of the universe is the measure of the interval between one position and the next immediately next to each of the primordial particles of space in its process of vibration, time is then a property of the particle which indicates the compass in which its motion vibration is happening, that is, the result of the relation between displacement and velocity, or

by Newton's equation of speed "t = d / v", since v for the primordial particles is a constant and equal to C (velocity of light), the time varies exclusively with the displacement in each period of vibration of primordial particles which in turn depends on the density of the space.

7 – Light is not a photon that travels through space, but rather as transported, a wave of energy that uses the substance space to propagate until it reaches a bulkhead where it discharges that energy by the particles of space in contact with the surface of the obstacle.

www.ingramcontent.com/pod-product-compliance
Lightning Source LLC
Chambersburg PA
CBHW040326220526
45473CB00009B/2582